GARGOURI Maroua
MARRAKCHI Chakib

"Viral hepatitis A: Diagnosis and management" guide

GARGOURI Maroua
MARRAKCHI Chakib

"Viral hepatitis A: Diagnosis and management" guide

Target audience: Interns, general practitioners, infectious diseases residents and family medicine residents

ScienciaScripts

Cover image: www.ingimage.com

This book is a translation from the original published under ISBN 978-620-6-72862-7.

Publisher:
Sciencia Scripts
is a trademark of
Dodo Books Indian Ocean Ltd. and OmniScriptum S.R.L publishing group

120 High Road, East Finchley, London, N2 9ED, United Kingdom
Str. Armeneasca 28/1, office 1, Chisinau MD-2012, Republic of Moldova, Europe
Printed at: see last page
ISBN: 978-620-3-61642-2

Pre-test :

MCQ1: Which viruses are preferentially hepatotropic?

A - Hepatitis A virus

B - Hepatitis C virus

C - Hepatitis B virus

D - Influenza virus

E - Varicella virus

Answers: ABC

MCQ 2: What signs are suggestive of acute hepatitis?

A - Icterus

B - Murphy's signC - Arthralgias

D - Right hypochondrium defence

E - Asthenia

Answers: ACE (symptoms)

MCQ 3: About the hepatitis A virus

A - It is an RNA virus

B - It is a virus that is resistant in the external environment

C - It is transmitted via the faecal-oral route

D - It is transmitted through the bloodstream

E - It is transmitted sexually

Answer: ABC (epidemiology + modes of transmission)

MCQ 4 What treatment should be started for acute common viral hepatitis?

A- Entecavir

B- Vitamin K

C- Paracetamol for pain relief

D- Interferon a2a

E- Symptomatic treatment

Response E (treatment)

MCQ 5: What method is used to confirm acute HAV?

A- Detect IgM antibodies to HAV in the blood.

B- Detect IgG antibodies to HAV in the blood.

C- Detect HAV-specific antigens in the blood.

D- Detection of HAV-specific surface proteins in blood.

E- Detect specific HAV RNA in the blood.

Answers: A (virological diagnosis)

Course outline

Introduction

Introduction :

Hepatitis A is an acute infectious disease of the liver caused by the hepatitis A virus. It is also known as "dirty hands disease" because it is most often transmitted through faecal-oral contact with contaminated food or water. Every year, around 10 million people worldwide are infected with the virus.

In emerging countries, and in regions where hygiene conditions are poor, the incidence of infection by the virus is close to 100%, and the disease is generally contracted in early childhood. Infection with the hepatitis A virus causes no detectable clinical signs or symptoms in over 90% of children, and because the infection confers lifelong immunity, the disease is not of particular importance to the indigenous population.

In other industrialised countries, on the other hand, infection is mainly contracted by non-immune young adults, most of whom are infected by the virus during travel to countries with a high incidence of the disease.

Hepatitis A does not present any risk of developing into a chronic form and does not cause chronic liver damage. After infection, the immune system produces antibodies against the hepatitis A virus, giving the patient immunity against future infections. The disease can be prevented by vaccination against hepatitis A, which has proved effective in controlling epidemic outbreaks throughout the world.

Epidemiology

Epidemiology :

The hepatitis A virus is highly resistant in the external environment. The reservoir is strictly human. It is estimated that 100,000 people are infected each year worldwide, and there are an average of 1,200 cases a year in France, around 40% of which are imported.

Human contamination occurs indirectly through the ingestion of contaminated water or food, or directly through contact with the faeces of an infected person (faecal peril).

In countries where hygiene conditions are good, circulation of the virus is low (seroprevalence

< 15%), small epidemics are the norm. This is the case for men who have sex with men (MSM), hence the recommendation to vaccinate this population.

If hygiene levels are low, children are infected before the age of 10, usually asymptomatically (seroprevalence is 70-100%). Since infection confers long-term immunity, epidemics are rare.

In countries where hygiene conditions are intermediate, infections occur later in life and are more often symptomatic (seroprevalence of 15-70%), and major epidemics can occur.

Geographical distribution zones can be characterised by their infection rate: low, medium or high. However, infection does not necessarily mean illness, since infants infected with the virus do not show any noticeable symptoms.

In low- and middle-income countries where sanitary conditions and hygiene practices are unsatisfactory, infection is common and most children (90%) are infected with HAV before the age of 10, most often asymptomatically. In high-income countries where health and hygiene conditions are good, infection rates are low. The disease can occur in adolescents and adults belonging to high-risk groups, such as injecting drug users, men who have sex with men, travellers to highly endemic areas and members of isolated populations (closed religious communities, for example).

In the United States, large-scale outbreaks have been reported among the homeless. In middle-income countries and regions with variable sanitary conditions, children often escape infection in infancy and reach adulthood without immunity.

Figure 1: Global distribution of hepatitis A (WHO 2012)

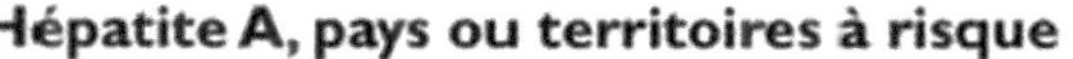

Transmission modes

Transmission modes :

The hepatitis A virus is transmitted mainly via the faecal-oral route, i.e. when an uninfected person ingests water or food contaminated by the faecal matter of an infected person. Within the family, this transmission can occur when an infected person prepares food for family members with dirty hands. Water-borne outbreaks, although rare, are generally associated with the use of contaminated waste water or inadequately treated water.

The virus can also be transmitted by close physical contact with an infected person (for example, during oral or anal sex), but it is not spread through normal person-to-person contact.

Virus de l'Hépatite A : épidémiologie

- **Transmission féco-orale**

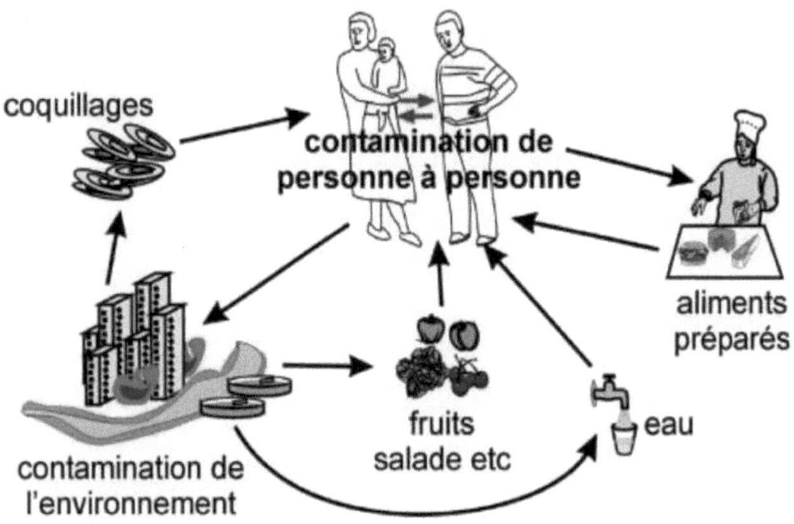

Pathophysiology

Pathophysiology :

In most cases, the virus enters the body orally and, because of its resistance to acidic conditions, reaches the intestinal villi after passing through the stomach. It appears that an initial phase of viral replication may take place in the intestinal cells, after which the virus reaches the liver, the virus' target organ, via the bloodstream. While no clinical symptoms are visible at this stage, viral replication increases in the hepatocytes and the virus is excreted in the bile, ending up in the faeces at concentrations of up to 109 per gram of faeces. At the end of this first asymptomatic phase lasting around 4 weeks, the virus is also found in the blood. The virus is not cytopathic on liver cells and modulates the host immune response, which explains the relatively long incubation phase. Once the infection has resolved, viral RNA may be found in the blood and faeces for 2 to 3 months, although this does not necessarily mean that infectious virus has been excreted.

Symptoms

Symptoms :

Symptoms of HAV infection, which are often absent or barely visible, appear 2 to 6 weeks after infection (incubation period). The patient is infected 2 weeks before symptoms appear and up to 2 weeks after they disappear.

In children under the age of 6, symptomless forms of hepatitis A are the most common
(70 %),

For older children and adults, the proportion of forms with symptoms increases
with age... as does the severity of hepatitis A.

The mortality rate among adults hospitalised for hepatitis A can exceed 1% in adults aged over 50.

Not very specific, the symptoms, when they exist, are varied and include 2 phases:

A pre-ictal phase lasting 1 to 3 weeks, marked by fever, fatigue, headaches, abdominal pain, nausea, diarrhoea, loss of appetite, joint and muscle pain and urticaria. These signs diminish and then disappear in the days following the onset of jaundice. An icterus phase (colloquially known as "jaundice") with the following symptoms: yellow skin and eyes, scanty, dark urine, discoloured stools and, rarely, pruritus.

SYMPTOMS OF

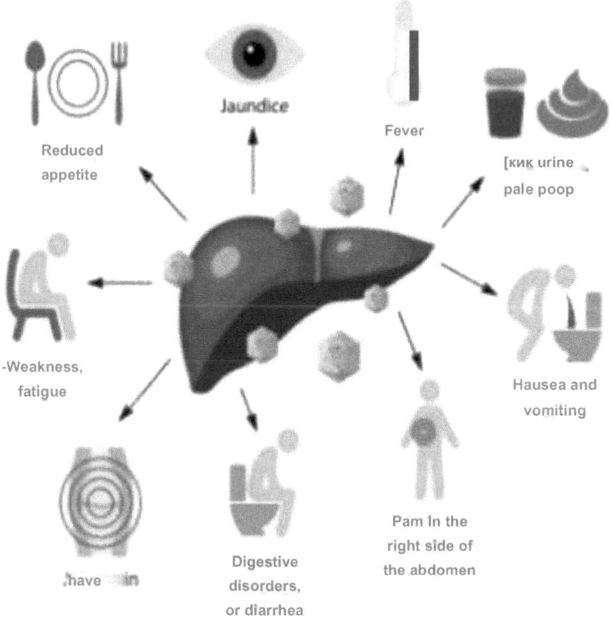

The deleterious effect of taking paracetamol at the same time should be emphasised, as it considerably increases the risk of fulminant hepatitis, especially when high doses are ingested. Fulminant hepatitis is rare.

Signs usually subside within a month, but fatigue may persist for up to 6 months. Hepatitis A never becomes chronic, but after the initial episode, a relapse of symptoms may occur in 3-20% of cases, but disappears without sequelae.

Additional examinations

Additional tests:

During the acute phase of the infection, other laboratory tests show :

- Constant hepatitis cytolysis: the damaged liver cells release their enzymatic content into the blood, especially alanine aminotransferase (ALAT), which can be found at very high levels (20 to 40 times the normal values) but which, in isolation, have no prognostic value.
- Bile retention in the blood: variable increase in bilirubin with a predominance of conjugated bilirubin.
- The prothrombin level should be systematically checked: a value of less than 40 or 50% could indicate a risk of acute liver failure (very rare, but very serious).

Imaging tests (liver, gall bladder, etc.) may be useful depending on the case (diagnostic doubt, co-morbidities, etc.).

Biological signs of seriousness: In addition to massive cytolysis and elevated bilirubin, coagulation disorders (PT < 50%, fall in factors II, V++, VII, IX), whether or not associated with DIC (thrombocytopenia and elevated D-dimer levels in addition to coagulation factor disorders) are the main factors with a poor prognosis, as is hyperphosphaemia.

The signs and symptoms of the disease appear more often in adults than in children.

The severity of the disease and fatal outcomes are greater in older age groups.

Infected children under the age of 6 usually show no noticeable symptoms, and only 10% develop jaundice.

Hepatitis A sometimes leads to relapses, i.e. a person who has just been cured falls ill again and has another acute episode, which will nevertheless lead to a cure.

Risk factors for hepatitis A

Risk factors for hepatitis A

Anyone who has never been vaccinated or previously infected can become infected with HAV. In areas where the virus is widespread (highly endemic), most cases occur in early childhood. Risk factors include the following:

1) Inadequate sanitation ;
2) Lack of personal hygiene
3) Lack of drinking water;
4) Staying in or travelling to endemic regions where hygiene conditions are poor, with inadequate or non-existent water purification systems,
5) Failure to vaccinate in a country where the virus is circulating,
6) Inadequate personal or collective hygiene measures,
7) Work at wastewater treatment plants.
8) Non-immunised people travelling to highly endemic areas.

Virological diagnosis

Virological diagnosis :

Virological diagnosis relies heavily on serology. During the acute phase, when cytolysis is linked to the antiviral immune response, the presence of anti-HAV IgM is highly specific for acute hepatitis A (Fig. 4). There are a number of key rules to be observed when indicating and interpreting this test. Testing for anti-HAV IgM should only be carried out in a clinical context suggestive of hepatitis; outside these circumstances, there is a significant risk of false positive reactivity. It should not be forgotten that many viruses induce hepatic cytolysis, which is usually moderate in the acute phase, and that polyclonal stimulation by certain viruses in the herpes group can induce a false reactivity in Ig detection tests. The risk of a false negative is low, but may occur at the very start of cytolysis (end of the prodromal phase), particularly when the patient is still febrile. In this case, IgM testing should be repeated on the following days (1-3 days). It should be noted that IgM can be detected during HAV vaccination in the weeks following injection; this should not be confused with acute hepatitis! The main purpose of testing for IgG or total anti-HAV antibodies is to check immune status. Positive serology, i.e. a titre >20 IU/mL, indicates either previous contact with the virus or immunity conferred by vaccination. This immunity is considered persistent and will protect against any HAV infection for the rest of the individual's life.

It is obviously possible to detect HAV RNA using RT-PCR techniques, either in the blood during the viremia phase, in the faeces or even in the liver at the time of cytolysis. This genomic research is of no use in routine medical diagnosis, but is useful for documenting the genotypic profile of a strain during an epidemic, or for detecting possible contamination of a food product or the environment. Assessing the level of viremia in an acute patient provides no useful information for clinical management. As part of the investigation of an epidemic, particularly in young children, it is possible to detect the antibodies secreted in saliva in order to determine the extent of the epidemic. This approach is less invasive than a blood test and, although less sensitive than a serum diagnosis, is sufficient to determine the number of cases affected.

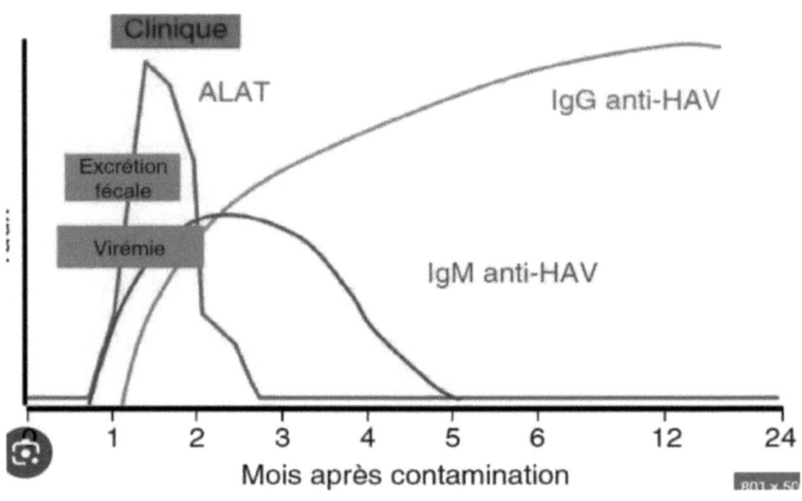
Clinique
ALAT
IgG anti-HAV
Excrétion fécale
Virémie
IgM anti-HAV
1
2
3
4
5
6
12
24
Mois après contamination

Treatment

Treatment

There is no specific treatment for hepatitis A. It can take several weeks or even months for symptoms to disappear.

It is important to avoid any unnecessary medication that may impair liver function, such as paracetamol.

In the absence of acute liver failure, hospitalisation is not necessary. Treatment is aimed at keeping the patient comfortable and maintaining an adequate nutritional balance, in particular by replacing fluid losses due to vomiting and diarrhoea.

Prevention

Prevention

Improved sanitation, food safety and vaccination are the most effective ways of combating hepatitis A.

Because the virus is excreted in large quantities in the faeces and is resistant in the environment, anyone infected should be informed immediately of the significant risk of transmitting the virus to those around them.

PREVENTION DÈ L'HEPATITE A

BONNE HYGIENE CORPORELLE

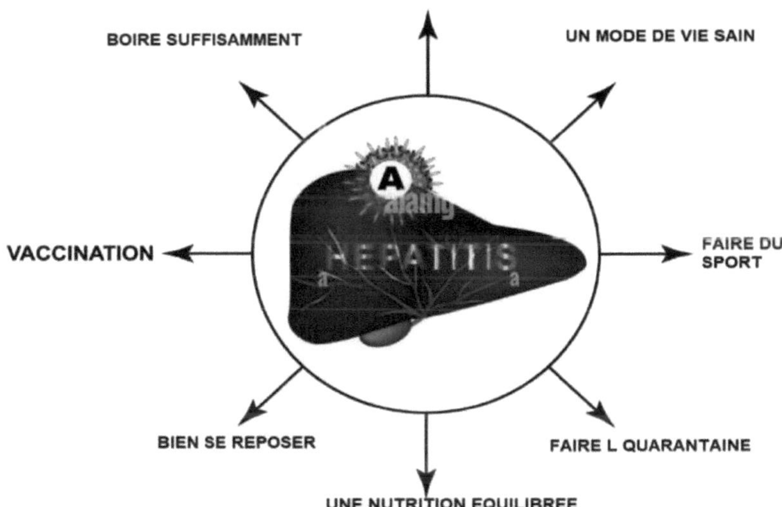

The first step will be to remind the infected person of the strict rules of hygiene in order to prevent secondary transmission.

The second measure will be to assess the need to vaccinate those around them. In fact, it is recommended that anyone in contact with a person with acute hepatitis A should be vaccinated as soon as possible (ideally within 14 days of the high-risk contact), without necessarily checking the serological status of those at risk. This measure has been shown to be highly effective in breaking transmission chains during localised epidemics. Particular attention will be paid to adults, especially the elderly, in order to limit the risk of severe hepatitis in these people.

Don't forget to report the case under the mandatory reporting scheme.

The hepatitis A vaccine was included in the Tunisian vaccination calendar from the 2018-2019 school year for pupils in the first year of basic education. The vaccine is made up of inactivated viruses from the HM175, GBM or CR 326F strains, depending on the manufacturer, and grown on MRC5 cells. The vaccination schedule is based on two injections, generally spaced 6 to 12 months apart. Vaccine efficacy is excellent (95%). To date, immunity appears to be persistent for life, and there is no recommendation for a subsequent booster. To check whether an individual has been immunised, an anti-HAV antibody test may be carried out. A positive titre indicates protection against infection.

Vaccination is recommended from the age of 1 for all travellers who are going to stay in a country where the level of hygiene is low, whatever the conditions of the stay. It is particularly recommended for people with chronic liver disease or cystic fibrosis.

A serological examination prior to vaccination (testing for total or IgG anti-HAV antibodies) is relevant for people with a history of jaundice, who have spent a prolonged period of time or their childhood in an endemic area, or who were born before 1945. The presence of anti-HAV antibodies (IgG) indicates previous immunity and does not justify the administration of vaccine doses.

Vaccination schedule

For adults and children aged 1 and over:

- 1st dose at least 15 days before departure
- 2nd dose (booster) 6 to 18 months later (depending on the speciality

chosen),

and then valid for life.

propagation of the virus :

- an adequate supply of drinking water;
- appropriate wastewater disposal in communities; and
- The application of personal hygiene practices, in particular regular

hand washing before meals and after using the toilet.

WHO action to prevent the spread of hepatitis A

WHO action to prevent the spread of hepatitis A :

WHO strategies guide the health sector in implementing targeted strategic actions to achieve the goals of ending AIDS, viral hepatitis (particularly chronic hepatitis B and C) and sexually transmitted infections by 2030.

These strategies recommend common measures and national actions targeting specific diseases, themselves supported by the action of the WHO and its partners. They take account of epidemiological, technological and trend developments in previous years, promote learning across the different diseases involved, and open up opportunities to harness innovations and create new knowledge to respond effectively to these diseases. They also call for intensified prevention, screening and treatment of viral hepatitis, focusing on the populations and communities most affected by and at risk of each disease, taking care to fill gaps and combat inequalities. They encourage synergies within the framework of universal health coverage and primary healthcare and contribute to achieving the objectives of the 2030 Agenda for Sustainable Development.

The WHO is organising World Hepatitis Day campaigns - one of its nine annual flagship campaigns - to raise awareness and understanding of viral hepatitis. For World Hepatitis Day 2023, WHO is focusing on the theme "One life, one liver" to illustrate the importance of the liver for a healthy life and the need to intensify prevention, screening and treatment of viral hepatitis to prevent liver disease and achieve the goal of eliminating hepatitis by 2030.

Key facts

- Hepatitis A is an inflammation of the liver that can progress from mild to severe.
- The hepatitis A virus (HAV) is transmitted by ingesting contaminated water or food or by direct contact with an infected person.
- Almost all people who contract hepatitis A recover completely, and are then immune for life. However, a very small proportion of people infected with HAV may die as a result of fulminant hepatitis.
- The risk of HAV infection is linked to the lack of drinking water and poor sanitation and hygiene conditions.
- There is a safe and effective vaccine to prevent hepatitis A.

Post-test :

1) **How is HAV transmitted?**

A - Vertical transmission from mother to child

B- Transmission through close physical contact with a

infected personC - Transmission by ingestion

contaminated food

D- Transmission by ingestion of contaminated water

E- Transmission by blood transfusion

Answer: BCD (transmission modes)

2) **What are the risk factors for HAV infection?**

A- lack of drinking water ;

B- sexual relations with a person suffering from acute hepatitis A ;

C- IV drug use

D- Corticosteroid treatment

E- movement of non-immunised people in highly endemic areas

ABE responses (risk factors for hepatitis A)

3) **Interpret this serological profile: IgM anti-HAV - , IgG anti-HAV +.**

A - Acute hepatitis A virus infection

B - Fulminant acute infection with the virus of

hepatitis A C - Long-standing infection with

hepatitis A virus

D - Chronic infection with

hepatitis A E - None of these

proposals

Response C (virological diagnosis)

4) **What are three preventive measures against HAV infection?**

A- An adequate supply of drinking water ;

B- Appropriate disposal of wastewater in communities

C- the application of personal hygiene practices

D- Inactivated injectable hepatitis A vaccines

E- Wearing long-sleeved clothing

ABCD answers (prevention)

References

1. Hepatitis A, World Health Organisation. https://www.who.int/fr/news-room/fact-sheets/detail/hepatitis-a

2. Hepatitis A virus (HAV) , Vincent Thibault. https://www.sfm-microbiologie.org/wp-content/uploads/2019/02/VirusHEPATITE-A.pdf

3. Hepatitis A , ***Sonal Kumar***, *MD, MPH, Weill Cornell Medical College Verified/Revised Jul 2024, MSD textbook consumer version https://www.msdmanuals.com/fr/accueil/troubles-du-foie-et-de-la- v%C3%A9ciliary-liver/h%C3%A9patitis/h%C3%A9patitis-a*

4. Understanding hepatitis A, health insurance. https://www.ameli.fr/assure/sante/themes/hepatite/comprendre-_hepatitis

5. Hepatitis A. https://www.elsan.care/fr/pathologie-et-treatment/infectious-and-tropical-diseases/hepatitis-a

Printed by Books on Demand GmbH, Norderstedt / Germany